D1402885

THE STREAM TEAM
ON PATROL

John G. Shepard

Published by Abdo & Daughters, 4940 Viking Dr., Suite 622, Edina, Minnesota 55435.

Library bound edition distributed by Rockbottom Books, Pentagon Tower, P.O. Box 36036, Minneapolis, Minnesota 55435.

Illustrated by Craig Parent

Edited by Deanne Rocquet

Library of Congress Cataloging-In-Publication Data

Shepard, John G.
  The stream team on patrol/written by John G. Shepard. p. cm. — (Target earth)
  Summary: Discusses the importance of streams and suggests ways to become
  involved in preserving this important part of the Earth's ecosystem.
  Includes bibliographical references and index.

  ISBN 1-56239-207-7

  1. Stream conservation—Citizen participation—Juvenile literature. 2. Water quality
  management—Citizen participation-Juvenile literature. [1. Steam ecology. 2. Water--
  Pollution. 3. Conservation of natural resources. 4. Pollution. 5. Ecology.]
    I. Title. II. Series.
    QH75.S4765 1993
    363. 73'946525--dc20   93-15386
                                   CIP
                                   AC

About the Author

John G. Shepard is an independent writer and president of Cascade Communications Inc., which designs experiential educational programs, videos, and curriculum materials to inform the public about environmental issues and motivate children to be effective stewards of our natural resources. John has an M.A. in cultural anthropology from Indiana University.

To Dad, my first and most loved river guide.

Thanks to the trees from which this recycled paper was first made.

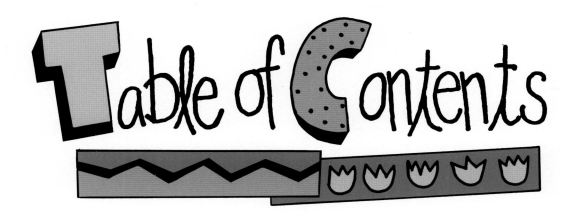

Table of Contents

Your mission, if you choose to accept it, is to go on patrol to rescue one of the most precious jewels in all the natural kingdom: a stream.

You will join a team—a Stream Team—that will be armed with special knowledge and keen observation skills. Working together, your team's assignment will be to explore and help save a stream that's under attack from a host of sly villains.

As your adventure unfolds, you'll learn how much more there is to the secret life of a stream than meets the untrained eye. You'll discover strange new life forms that you might never have imagined in your wildest dreams. Most importantly, you'll make a real difference for thousands of creatures—including the two-legged ones—whose lives are involved with the stream in many fascinating ways.

Chapter 1

Operation Save a Stream

Streams Under Attack

Streams are in serious trouble all across the country. Pesticides, fertilizers, chemicals from factories—even the soil itself—is washing into the water, destroy conditions needed by fish, insects, and other aquatic (underwater) animals for survival. Alien creatures and plants that find their way into streams from distant places upset important relationships between the stream's many inhabitants. Dams and other human constructions often so drastically change natural conditions in streams that many plants and animals just can't cope.

Humans have caused major damage to streams even though, like the other creatures on the planet, we depend on them for our very survival. Most people aren't aware that worldwide, 90 percent of our drinking water comes from streams and rivers. That's a lot of water; but we use ten times that much water from rivers and streams to irrigate crops, for industry, and for creating electric power. Once we've taken the water they bring us, we often show our thanks by using streams and rivers as dumps for our waste.

North America's watery highways weren't always sewers. Before the invention of planes, cars, and trains, boating on rivers and streams was one of the main ways people traveled about. Many of our biggest cities grew beside rivers because they provided transportation, power for factories and electricity, and water for drinking. As we've become more and more dependent on railroads, freeways, and airplanes, however, we've forgotten about the waterways that truly are the lifeblood of our continent.

The good news is that even though the problems facing streams are serious, everyone can do something to improve them—and that includes you. Doubt it? Consider the following true story.

A group of students in Illinois were testing the water quality of a stinky stream that ran through a historic town that tourists love to visit. The students found that the creek's smell was caused by the fact that the town had no sewer system. The lovely historic homes were dumping raw sewage into a stream played in by children. The students double-checked their water quality test results and wrote letters to their U.S. and state senators and to several government agencies. **The result:** the town was given three years to build a decent sewer system and clean up its stream.

This kind of work isn't easy. It even has its risks. But there's no work more important, and you can take real pride in the effort you make on behalf of your adopted stream. So, if it's for you, create or join a Stream Team and, with your teammates, take the Stream Team pledge.

Stream Team Pledge

For the gift of fresh water I am thankful. On behalf of all the other creatures who share this gift with me, I accept the challenge of caring for a stream and the web of life it supports.

Activities

1. Form a Stream Team.

   What? There is no Stream Team in your school? Then your first task is to start one. Share this book and the idea of starting a Stream Team with your friends. Once you get a team together, find an adult advisor who would like to work with your group. The adult's job is to assist with planning activities and help keep an eye on safety when you're on patrol. Remember, though, the Stream Team is YOUR team; it's up to you to plan and carry out your mission.

2. Make it official: The Stream Team T.

Needed materials: T-shirts, permanent paint or markers, paper, and scissors.

Before going on patrol, your team will need to get into uniform. Using the Stream Team logo in this book, make your team's own T-shirts by tracing stencils and painting the design onto a shirt with permanent paints or markers. (Another option is to explore making silk-screened T-shirts with your art teacher. For more information, see the Target Earth Earthmobile book *Eco-Arts & Crafts.)*

Chapter 2

Basic Training I:
A Stream and Its Watershed

There's no getting away from streams. They are everywhere. Wherever you happen to be at this or any other moment, the rain that falls to the ground around you will flow downhill toward the nearest stream. The entire area of land that drains into the stream is called its basin, or watershed.

An effective Stream Team member understands that a stream's watershed—where it is, how big it is, and most importantly, what's happening within it—is very important. A stream's health depends on the quality of the water that feeds it, and that water flows across the land within the stream's watershed. Caring properly for the land is often the key to improving the health of the stream that winds through it.

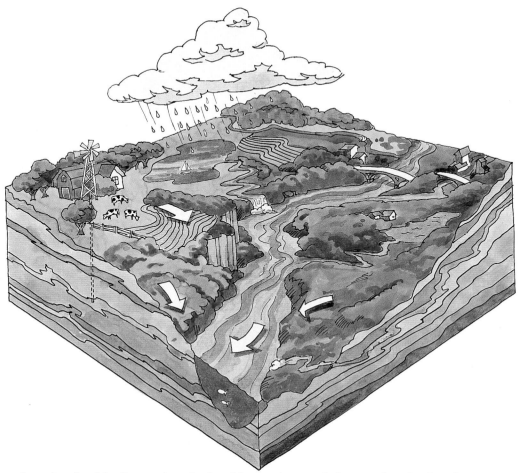

A watershed is the region drained by a stream, lake or other body of water.

Activities

1. Stream search.

You will want to find a nearby stream that's small enough for safe wading and at least partly in an area with houses, farms, factories, or other human developments that are likely to affect the stream. Call local conservation organizations and state agencies that take care of rivers and streams to get information about streams in your area that you could adopt.

Request any maps and surveys of streams that interest you. Contact the Izaak Walton League (1401 Wilson Blvd., Level B, Arlington, VA 22209; telephone, 703-528-1818) and request a copy of their *Save Our Streams Kit.* The kit has good background information and important details about several activities suggested in this book. Based on this information, choose the stream you'll adopt.

2. Mapping your watershed.

Needed materials: a compass, pencil, U.S. Geological Survey (USGS) topographical map of your stream and the area around it, good photocopy of that map, and any other maps you've been able to get of your stream.

A USGS topographical (or "topo") map, with contour lines showing the shape of the land, is a fascinating way to study the surface of the Earth. A local outdoor store may have the topo map for your stream's watershed. If not, you can order an index of USGS maps for your state by calling the USGS toll-free at 1-800-USA-MAPS. The index will indicate which map or maps you'll need (your stream's watershed may extend onto two or more maps). For help understanding how topo

A USGS topographical map of Forest Lake, Minnesota.

maps show the changing slope of the land by means of contour lines, you can check out the book, *Basic Essentials of Map and Compass* (see Connect With Books on page 38) from your library.

It's time to get out in the field face-to-face with the stream of your choice and its watershed. Take your maps and compass to a nearby hilltop from which you can see your stream's valley (get permission before entering private property). Place the topographical map beside your stream map on the ground so that the top of both maps are pointing north. Compare the changing slope of the land around you with the hills and valleys outlined on your topo map. See if you can trace the course of the stream through the valley.

Next, find the boundaries of your stream's watershed and draw these onto the photocopy of your map. Notice how the colored shading on the original topo map shows where trees are growing. Draw the boundaries of any farms, housing developments, fields or areas that have been cleared of trees. You will probably need to go to several spots to fill in the details of the entire watershed. (Note: If your stream's watershed is too big to map by going to several spots, just map the half-mile strip of land along both sides of the stream and along both sides of any creeks that feed your stream.)

You now have a map that tells you many important things about your stream's watershed. It will be an important tool that you'll use when you go on patrol. But before heading out, we'll take a peek beneath the surface of the water to explore a few stream secrets.

Chapter 3

Basic Training II:
Secrets in the Life of a Stream

A stream is like an anthill, only more so. Watching the ants march in and out of the opening to their home gives you no clue that an amazing spaghetti-pile network of tunnels lies just below ground. In the same way, a single fish rising to the surface of a stream to snatch a bug for dinner doesn't even come close to suggesting all the strange and wonderful things going on within and around the stream.

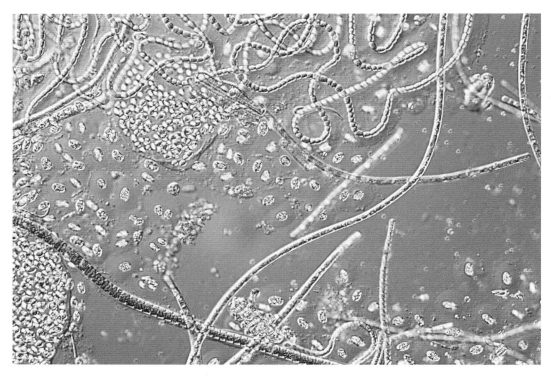

A microscopic photograph of blue algae. Algae is at the bottom of the food chain.

One way to make sense of all the activity taking place in a stream is to look very carefully at who is eating whom. This view shows an interesting pattern that ecologists call a food chain. The basic idea is that energy from the sun is used by plants and animals and, when they are eaten, passed on to animals higher on the food chain.

At the bottom of the food chain, tiny plantlike creatures called algae (pronounced "AL-gee") form a soft, slimy green layer you can see and feel on underwater rocks and logs. Algae need direct sunlight to grow, so you'll find more of it in large unshaded streams or in smaller wooded streams during the spring before budding leaves overhead create too much shade.

Though algae may not look like dinner to us, snails, clamlike creatures called limpets, and some insects that begin their lives

14

underwater consider it a treat. As the algae is eaten, some of the energy it received from the sun is passed on to these creatures, and so to the next level of the food chain.

Leaves, pine needles, flower petals, and bark dropped into the stream by trees and plants (which use the sun's energy to grow) form an even more important part of the bottom link of the food chain.

This is especially true for small streams in wooded areas. These pieces of plant matter collect at the bottom of the stream where millions of microscopic bacteria, molds, and fungi devour them. The result is the same as what happens to a dead tree lying on the forest floor: the plant matter in the stream slowly falls to pieces, or decomposes. The bacteria, molds, and fungi that do the work are called decomposers.

A group of underwater animals called shredders, including snails and insects, help the decomposers by breaking leaves into small chunks. In the process these shredders also feed on the decomposers themselves. This places them one link up from the bottom of the food chain. Since they consume, or eat, the decomposers, shredders are called primary (or "first-level") consumers.

The sun's energy, which has already passed through the algae and plant matter to the decomposers and shredders, now reaches the next level of the food chain. Animals that feed on the primary consumers make up this higher level. These small fish and larger insects are called secondary consumers. There are also third-level consumers—larger fish, small land animals like otter and mink, waterfowl, and humans with fishing poles, all of whom feed on the secondary consumers. Again, the energy received by the third-level consumers when they eat the secondary consumers is the same energy originally provided to the algae and plants by the sun.

A food chain is the transfer of food energy through a series of organisms
like birds, snakes, frogs, and insects. Each organism eats the member below it.
The sequence begins with algae and green plants (producers), herbivores (primary
consumers), and carnivores (secondary consumers). Decomposers
act at each stage and at the end of the chain.

The idea of a food chain helps make sense of the complicated world in and around a stream. But if you spend enough time exploring that world you'll discover not every relationship between a consumer and its dinner fits neatly into the concept of a food chain.

For instance, decomposers munch on dead fish as well as leaves—so you could think of them as belonging at the bottom of the food chain or at the top. Some aquatic animals eat both vegetation and other animals; where do they fit? There's even a type of beetle common to many streams that eats tiny fish—and when the surviving fish get big enough, they eat the beetles.

In fact, you could think of a healthy stream as having many food chains connected to each other in different ways. If you traced the lines between all these food chains, the result would be a crisscross of threads that looks like a spider's web—so much so that ecologists call a stream's connected food chains a food web.

Exploring your stream, you also might discover that the number of individual animals decreases as you go up the levels of a food chain. The reason has to do with the fact that most of the sun's energy is lost as it's passed from link to link. At each level of a food chain the plants and animals use 90 percent of the energy they receive simply to grow and live. Ecologists, who can actually measure the food energy available at each level of a food chain, show this flow of energy as a pyramid. The animals at each level can get only ten percent of the energy available one level down. This explains why food chains rarely have more than three or four levels; there just isn't enough energy available at the top to support another group of hungry animals.

A food chain (short red arrows) is often tied to other food chains to form a complex food web where organisms from different food chains feed off each other.

When you understand how the food web works, you can begin to see the stream as a system of living things, an ecosystem, in which the needs of different types of plants and animals influence each other. Just as a pebble dropped into a pond sends ripples across the surface, a healthy stream's food web constantly changes as the numbers of plants and animals living within it go up and down.

For example, if changes in water temperature causes the total number of primary consumers to get too small, some of the secondary consumers won't have enough to eat and gradually their population might shrink too. As this ripple effect spreads to the third-level consumers, their numbers may change as well.

Some naturally occurring events like forest fires, floods, or droughts can cause sudden changes throughout a stream's ecosystem. Every now and then even an entire species of animals may become extinct in a stream due to natural causes.

But most natural changes in a stream's ecosystem happen slowly and extinctions are rare. As the next chapter shows, it is usually human activities that bring about sudden and far-reaching changes that make life very difficult for many stream creatures.

Activities

1. Stream critters.

Call nature centers in your area and make arrangements to visit any that have a naturalist who can introduce your team to the creatures that live in and around a stream. Ask the naturalist what plants and animals make up the food web in healthy streams in your area. Take notes of the information he or she gives you.

2. Build a food web and energy-flow pyramid.

Materials needed: Paper and pencil.

Based on the information you get from the naturalist about the creatures and plants that are found in healthy streams in your area, build models of what might be in your stream's food web and an energy-flow pyramid. Make the food web by building a series of food chains and connecting them wherever animals eat or are eaten by creatures in other chains. Using the diagram on this page as a guide, fill in the names of the aquatic plants and animals in your area that belong to each level of the pyramid. Some may not fit neatly in one level—you can list them more than once or choose which level seems most important.

Third-Level Consumers

Secondary Consumers

Primary Consumers

Producers

Chapter 4

Basic Training III:
Know the Enemy

What comes to mind when you think of the problems humans have created for streams? If you're like many people, you may picture an ugly pipe from a factory dumping foul-smelling and strangely colored chemicals into the water. The factory is run by a nasty thug who only cares about making money. The poor fish and other animals living downstream have become glowing three-eyed mutants from the polluted water.

There may be a few factory owners who fit that image, and mutant fish, though perhaps not three-eyed and glowing, do exist. But this kind of pollution is relatively rare. More often, serious problems are created for streams by ordinary people and their normal, day-to-day activities. And the real villains—things like lawn fertilizers, garbage, and soil washing into the stream—can sometimes be hard to track down. We'll take a quick look at each of the main suspects one at a time.

Point Source Pollution

Point source pollution is pollution that comes in strong doses from one or a few sources. Besides factory discharge pipes, you might spot point source pollution at sewage treatment centers that can't properly manage all the waste they receive. Leaking chemical

storage tanks, landfills near streams, chemical dumps, and chemical spills from trains, trucks, or pipes are other possible sources. Often point source pollution involves materials that are toxic, or poisonous. Exposed plants, animals, and humans can become sick, have problems producing healthy offspring, or die.

Non-Point Source Pollution

Non-point source pollution enters the stream in small amounts from many different sources. Fertilizers made to help grass and plants grow on lawns, gardens, golf courses, and farms cause problems when they accomplish the same results with algae and underwater plants in streams. Large growths of algae and aquatic plants shut out sunlight and change the chemical make-up of the stream, which makes life tough for aquatic animals. Non-point source pollution can also include leaking household waste disposal systems, dumped garbage, pesticides and herbicides, and animal waste where livestock are allowed to graze in or near streams. Drains carrying rain water from city streets can also pollute streams with oil and gas leaked by cars.

Thermal Pollution

Most aquatic creatures can't take the heat when hot water is added to streams by industries along their banks. This problem can occur near power plants and other industries that use water to cool super-hot pipes.

Siltation

Even the earth itself can be trouble for stream life. Too much soil can wash into streams when nearby trees and vegetation are cleared for logging, farming, highway construction, and housing developments. The soil can cover underwater gravel beds that are needed by insects and used by fish to spawn, or lay eggs. It also can kill fish eggs, damage fish gills, make hunting difficult for consumers, cause flooding, and even completely fill stream channels.

Structural Damage

Structural damage is damage caused by changes to a stream's flow or channel. Humans can be busier than beavers in changing streams to control flooding or to use the water for other purposes.

Dams, which turn streams into lakes, disrupt conditions for aquatic animals living upstream. Water levels and temperatures downstream of dams can change wildly because of changing flow rates. Dams also can prevent fish from traveling upstream to spawn. Levees (walls built along stream banks) and efforts to deepen and straighten stream beds to control floods keep fish from reaching important marshy areas, or wetlands, along the stream's banks. These structural changes cause wetlands, which are also very important for waterfowl, to dry up.

Exotic Species

A stream's ecosystem can be highly disturbed when alien creatures enter the stream. Sometimes these exotic (or "not-native") plants and animals are introduced on purpose to "improve" the stream; for example, a new kind of fish may be added to improve fishing. Other times they enter the stream by accident, such as when a boat carries on its hull plants or small animals from distant rivers or lakes. In either case, the results can be disastrous for animals that have developed no defenses against these intruders.

Activities

1. Pollutants at home.

With your parents' help, find and list any pollutants in your own home that can damage streams. Find out how your family uses and disposes of old paint, batteries, glass and plastic bottles, cans, newspapers, food and yard wastes, solvents, detergents, used motor oil, insecticides, pesticides, and fertilizers. Call the local office of your state extension service to find out if there are alternatives for using and disposing of any of these materials that are toxic. You can discuss with your parents starting a recycling system for cans, bottles and newspapers and a backyard compost for food and yard waste.

2. Spread the word.

Share what you've learned about proper use and disposal of household pollutants with the rest of your school. Make posters that tell other students about the problem (you may want to involve your school's art department) and hand out information on how their families can best take care of these materials. Write letters to the editor of your community newspaper and share the same information.

3. Drain strain.

The next time it rains, notice where the rainwater goes that drains from your yard. If it disappears into a storm sewer on your street, call your city's department of public works to find out what body of water it eventually enters and whether it's filtered or treated before being discharged. If you live in a rural area, trace the flow of any drainage ditches from your yard to the nearest stream or body of water. If you use chemical fertilizers or pesticides on your lawn, gardens, or fields, you can explore with your parents the idea of switching to organic materials or planting vegetation to help filter these chemicals before they leave your land.

Chapter 5

The Stream Team
On Patrol

Now that you've been introduced to the life of a stream and to the sorts of problems it may have, it's time to take a close look at your adopted stream to find out how it's doing. In this work you'll be acting like a team of doctors, taking the pulse of your stream's health and making plans for improving it.

Activities

1. Stream walk.
Based on the information you've gathered about your stream and its watershed, choose a section that's one to two miles long for your stream walk. A section that flows through woods or an undeveloped area and past farms, housing, or some kind of development would be best. Here's some things you'll need for your outing.
• Clothing: your Stream Team T-shirt, of course, plus pants or shorts for wading, gloves, tennis shoes or boots that won't be hurt if they get wet, sun hat or sun block, and a dry change of clothes in a water-tight plastic bag.

• Equipment: notebook and pencil for everyone, first aid kit, the stream quality survey instructions from your Save Our Streams kit (see Connect With Books to order), a window screen without holes (called a kick seine—instructions in SOS kit) and some plastic jars for collecting insects, a powerful magnifying glass, topographic and stream maps, compass, camera and film, and water for drinking.

• Preparation: Contact land owners for permission if your stream section passes through private property. Ask a naturalist or fresh water biologist to join you (this is important for your first stream walk, as identifying the insects you collect can be difficult at first).

Recommended Stream Team clothing and equipment.

Stream Walk Safety

- Adult advisor should be with you.
- Look out for your buddy at all times.
- Wade in water no deeper than your knees.
- Beware of drop-offs.
- Watch out for slippery rocks, trees, and sharp objects.
- Stay on shore if the water shows signs of toxic pollution: colored or oily film on surface, strong smell, or large areas of white foam more than three inches deep. Avoid discharge pipes from factories or sewage treatment plants and sealed containers that could hold toxic materials. Contact pollution control agencies and tell them of these conditions.
- Wash hands afterward.

Begin walking upstream from the downstream end of your section so that underwater rocks and silt disturbed by your feet won't spread to areas you haven't explored yet. As you walk, look for and take note of the following:

1. Note on your map the locations of creeks entering the stream; road crossings; farms, homes, factories, and other development; dams and structural changes; stream banks without vegetation where soil is washing into the stream; discharge pipes and pipes that drain rain water from farm fields; and riffles (small rapids).

2. Note any changes in the color or smell of the water (refer to the table of water colors and odors in SOS kit) and record where these occur. Also note where garbage has been dumped or washed downstream and, in general, how much litter is found in each part of your stream.

3. Look for, list, and note the location of wildlife in and around stream, including fish, birds, and land animals. Note surroundings of any wildlife sightings.

4. To measure the health of your stream, find several areas about a quarter-mile apart with riffles 3-12 inches deep. Use these sites for collecting insect samples and studying them according to the Stream Quality Survey Instructions in the SOS kit. Choose at least one site that is in a wooded or undeveloped area and at least one downstream of farms or housing developments. Complete the SOS Stream Quality Survey forms. Note: be sure to carefully note the location of your test sites so you can find them later for further testing.

5. Once you've completed your stream walk, drive with your adult advisor to several other places along the stream above and below your adopted section to do additional stream quality surveys. Information from these surveys can be added to surveys taken during your stream walk to give you a more complete picture of your stream's health.

6. Stream rescue plan. Now your team has the information you need to make a diagnosis and put together a plan to restore your stream to better health. It's very important to find out exactly what's ailing your stream and why so that you can prescribe the right treatment for its problems. This can be tricky ecological detective work. The information in your SOS kit will help you understand what kinds of pollution or human activity may be causing poor stream quality. But you'll have to track down the actual sources of those problems within the watershed.

WATERSHED MAPS CAN PROVIDE USEFUL INFORMATION, INCLUDING HUMAN ACTIVITIES THAT AFFECT YOUR STREAM.

SO THE STREAM TEAM WENT ON PATROL TO SEE FOR THEMSELVES WHAT SHAPE THEIR STREAM WAS IN AND WHAT THEY COULD DO ABOUT IT.

First, list on separate sheets of paper the location and results of each of your stream survey sites. At each site where water quality was poor, list likely causes of the poor water quality according to the information in your SOS kit.

Then look for the villains that might be causing problems for your stream by reviewing your notes of the conditions that you noticed during your stream walk and nearby developments on the watershed map. Remember: pollution, like water, flows downstream, so you should always be looking upstream to find causes of problems. (Note: If you didn't map the entire watershed because of its size you may want to do more mapping now in areas where your survey results show poor water quality and you can't find a cause for the problem.)

Once you have some likely suspects for each problem, the next step is to plan solutions. For instance, if you noted garbage dump sites or lots of litter along the stream, you can follow the guidelines in your SOS kit and organize a cleanup event and invite other groups to join you from your school and community. Write letters to the editor of your community newspaper about the problem. Contact community officials and urge them to post no litter signs and to enforce dumping laws better.

Elsewhere, you may have noticed muddy water in a part of the stream where the banks had no vegetation. You could include in your plan a project to get permission from land owners to plant vegetation along the stream banks to keep soil from washing into the stream.

Read through your SOS kit and other resources, such as the *Adopting A Stream: A Northwest Handbook* (see Connect With Books on page 38), to help form ideas for developing and carrying out your plans.

Whatever plans you make, you should continue monitoring the water quality in your stream about six times each year at the same survey sites used in your stream walk. This will help you keep track of your stream's health over time and will tell you which of your rescue projects are effective. See the SOS kit for information on continued monitoring.

Chapter 6

Taking It to the Streets

A Lesson from the Monkeys

Some scientists studying a group of potato-eating monkeys on a Japanese island during the 1950s noticed something interesting. Imo, a young female monkey named by the scientists, discovered that she could clean the sand and grit from her potatoes by washing them in the ocean. The other young monkeys soon caught on and began washing their potatoes, too. It took a while, but finally even the grown-ups started taking notice. Within a few years all the monkeys on the island, inspired by Imo's good idea, were washing their potatoes.

What's unusual about this story is that on Imo's island young monkeys learned their eating habits from their parents—except in the case of clever little Imo. People, perhaps, aren't so different from monkeys.

This story is a good one for your Stream Team to keep in mind. In order for our streams to have a healthy future, it will take changes in daily habits and behaviors that we all take for granted. The trick is to get enough people—especially adults—to be careful about things like using lawn fertilizers and controlling erosion so that ecologically healthy practices become the accepted way of doing things.

Remember that you've done a great deal already to help bring about this change for your adopted stream. By sharing with a wider audience what you've learned about taking care of your stream you just might do for your community what Imo was able to do for her island.

Activities

1. Storm sewer labeling.

Materials needed: Heavy-duty paint, brushes, and stencils.

Gas and oil leaked from vehicles onto street surfaces where it is washed into storm sewers by rain water is a problem. But an even bigger concern is people dumping used motor oil, anti-freeze and other chemicals into storm sewers. A great way to reduce dumping of these materials is to label all storm sewers in your community with a painted warning like the one on page 35.

Note: Your team will need to get permission from the city to do this project. Your adult advisor can help you contact the right office.

2. Start an Adopt-a-Stream campaign.

Coordinate community participation in getting your entire stream or other streams in your area adopted and cared for. The campaign might involve developing a brochure that describes your stream's features and problems and distributing it to people who live nearby. You can give presentations to youth organizations and citizen groups inviting them to join you in keeping the stream healthy (use pictures taken of your Stream Team activities in your presentation). You can also write letters to the editor of your community newspaper inviting citizens to join your campaign.

3. Go on stage.

Write and produce a play or make a video about your stream's problems. Perform the play or show the video to others in your school or community. Use the Stream Team comic strip in this book for the basis of your play.

4. Learn the law.

Studying the laws that protect your stream from pollution will help you understand what actions you can take if you find polluters who are damaging the stream. Conservation organizations in your community that work to stop polluters may be able to help you.

5. Celebrate your stream and its team.

Celebrating the importance of your stream and the good work that your team and the community has done to save it is an important part of your Stream Team mission. Working with other groups who are concerned about your stream, you can help organize a stream festival, perhaps at a stream-side park. Invite your school to participate in the event.

By now your stream will have benefited greatly from the dedicated work of your Stream Team and the community. As you've probably learned, taking care of planet Earth is a job that's as fun and satisfying as it is challenging. Taking care of the lands and waters is something everyone can help with, no matter where you live and how old you are.

Congratulations! And keep up the good work!!!

ONCE YOU UNDERSTAND YOUR STREAM AND ITS PROBLEMS, LET THE WORLD KNOW. ORGANIZED CLEAN-UPS AND STORM SEWER STENCILING ARE TWO PROJECTS THAT CAN MAKE A DIFFERENCE FOR YOUR STREAM.

CONGRATULATIONS! ALL YOUR HARD WORK HAS MADE A BRIGHTER FUTURE FOR YOUR COMMUNITY AND STREAM.

Glossary

Algae—Tiny plants that lack stems, roots, and leaves. Usually found in water.

Carnivore—An animal that eats the flesh of another animal.

Decomposers—Bacteria, molds, and fungi that devour plant matter.

Drought—A long, dry period of weather, with little or no rain.

Ecosystem—The interaction of plants, animals, and other natural elements in an interrelated system.

Exotic species—Plants or animals that are introduced to an ecosystem from distant places and create a new set of problems.

Food chain—The flow of nutrients and energy among a series of organisms that feed on each other.

Food web—A connecting series of food chains.

Herbivore—An animal that feeds on living vegetation.

Izaak Walton League—A national environmental organization in Virginia whose members protect America's soil, air, woods, water, and wildlife.

Kick seine—A large net used to catch aquatic insects.

Levee—A dike made of earth, stone, or concrete. A levee is built along the banks of a river to help control flooding.

Limpet—A clamlike creature found clinging to rocks along marine shores. It has a caplike shell and large muscular foot.

Non-point source pollution—Pollution that comes in small amounts from many different sources.

Organism—A living plant or animal.

Pesticide—A chemical substance used to destroy bugs and other pests.

Point source pollution—Pollution that comes in strong doses from a single source, like a factory discharge pipe.

Primary consumers—Animals in a food chain that eat plants. Shredders and herbivores are primary consumers.

Riffles—Small rapids in a stream or river.

Secondary consumers—Animals that feed on primary consumers. Small fish and larger insects are secondary consumers.

Shredders—A small group of underwater animals, including snails and insects, that eat decomposers.

Siltation—The filling up of a stream or reservoir with sediment.

SOS Kit—Save Our Streams Kit developed by the Izaak Walton League. The kit has good background information and important details about Stream Team activities.

Thermal pollution—An increase or decrease in water temperature that damages organisms in lakes, rivers, streams, and other bodies of water.

Third-level consumers—Animals that feed on secondary consumers. Larger fish, small land animals, and humans are third-level consumers.

Topographical map—A map with contour lines that shows the surface features of a place or region.

Watershed—The total area drained by a stream or river.

Connect With Books

Adopting A Stream: A Northwest Handbook. Steve Yates, 1988, University of Washington Press.

Basic Essentials of Map and Compass. Cliff Jacobson, 1988, ICS Books.

Rivers, Ponds and Lakes. Anita Ganeri, 1991, Dillon Press.

Save Our Streams Kit. Izaak Walton League, 1401 Wilson Boulevard, Level B, Arlington, VA 22209 (a small fee is charged for the kit).

Index

TARGET EARTH COMMITMENT

At Target, we're committed to the environment. We show this commitment not only through our own internal efforts but also through the programs we sponsor in the communities where we do business.

Our commitment to children and the environment began when we became the Founding International Sponsor for Kids for Saving Earth, a nonprofit environmental organization for kids. We helped launch the program in 1989 and supported its growth to three-quarters of a million club members in just three years.

Our commitment to children's environmental education led to the development of an environmental curriculum called Target Earth™, aimed at getting kids involved in their education and in their world.

In addition, we worked with Abdo & Daughters Publishing to develop the Target Earth Earthmobile, an environmental science library on wheels that can be used in libraries, or rolled from classroom to classroom.

Target believes that the children are our future and the future of our planet. Through education, they will save the world!

Minneapolis-based Target Stores is an upscale discount department store chain of 517 stores in 33 states coast-to-coast, and is the largest division of Dayton Hudson Corporation, one of the nation's leading retailers.